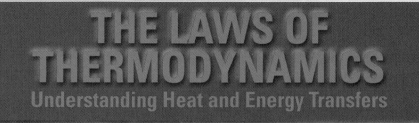

# The Library of Physics™

# THE LAWS OF THERMODYNAMICS

## Understanding Heat and Energy Transfers

**Rose McCarthy**

The Rosen Publishing Group, Inc., New York

Published in 2005 by The Rosen Publishing Group, Inc.
29 East 21st Street, New York, NY 10010

**Library of Congress Cataloging-in-Publication Data**

McCarthy, Rose.
The laws of thermodynamics : understanding heat and energy transfers / by Rose McCarthy.—1st ed.
    p. cm. — (The library of physics)
Includes bibliographical references.
ISBN 1-4042-0336-2 (library binding)
1. Thermodynamics.
I. Title. II. Series: Library of physics (Rosen Publishing Group)
QC311.M33 2005
536'.71—dc22

2004019127

*Manufactured in the United States of America*

**On the cover:** A view of melting ice

# Contents

# Introduction

Thermodynamics is the branch of physical science that deals with the transformations of energy. The word "thermodynamics" is derived from the Greek *therm*, or heat, and *dynamis*, or force. The term originally referred only to the study of heat, but its scope expanded as scientists learned more about the nature of both heat and work. Study into thermodynamics began in the early nineteenth century when scientists began examining heat and its relation to energy. Today, the field of thermodynamics is important in engineering and many branches of science, including physics, chemistry, geology, biology, and others.

Thermodynamics is based on a few basic postulates known as the laws of thermodynamics. They were established through experimentation and scientists' experience with the behavior of energy. The second law of thermodynamics was recognized in the nineteenth century before any of the others, followed shortly by the first law. The third and finally the zeroth laws were added in the twentieth century.

The first law of thermodynamics is the law of conservation of energy. Energy can be neither created nor destroyed, although it can be transferred from one form to another. The third law deals with the behavior of a property called entropy at very low temperatures and the measurement of entropy. The zeroth law clarifies the definition of temperature.

Thermodynamics is in force in the operation of a hot air balloon. Lifting the balloon is achieved by increasing the temperature inside the balloon. The balloon is lowered when the temperature is allowed to cool.

The second law of thermodynamics is one of the most fundamental rules of physics. While the first law clarifies the nature of energy, the second places limits on the direction of energy transfers.

There have been many different statements of the second law. One formulation states that when heat is transformed into work, it is impossible to transform all of the heat into work without discarding some heat. Another formulation states that heat can flow only from a hotter object to a cooler object, never the other way around.

Implications of the second law of thermodynamics led one nineteenth-century physicist, Rudolph Clausius, to coin the word "entropy." Entropy is a measure of the tendency of energy to move outward from a concentrated area. Clausius formulated another statement of the second law based on his understanding of entropy: the entropy of the universe tends toward a maximum. Natural processes tend to result in an increase in entropy. The very concept of entropy has profound implications in science, ranging from the behavior of subatomic particles to the future of the universe.

# 1
# The First Law of Thermodynamics

The first law of thermodynamics is often called the law of conservation of energy. The first law states that energy can neither be created from nothing nor destroyed. For every physical process in which energy is transferred—boiling water, starting a car, turning a waterwheel—the total amount of energy involved never changes.

In formulating the first law, scientists redefined the concept of energy. They realized that heat and work were two different ways of transferring energy, and that each could be converted into the other. We take this for granted in our daily lives today. Coal or gas burnt at power plants produces heat, which in turn provides the energy that runs our refrigerators, computers, and dishwashers. But to scientists of centuries past, discoveries about the nature of energy changed the fields of physics and engineering.

## The Nature of Heat

Thermodynamic analyses often focus on processes within an area of interest called a system. A thermodynamic system is defined by its boundary, which

This power plant in New Roads, Louisiana, is fueled by coal. The power plant transfers heat into electricity by burning the coal, which generates electricity. Most of the electricity in the United States is produced this way.

separates it from the surroundings. A system can be a glass of water, an engine, or an amount of gas in an airtight cylinder. Some systems may interact in some way with the surroundings. An isolated system, on the other hand, is a type of system that does not exchange heat or work across the boundary. Scientists work with many different kinds of systems to understand thermodynamic interactions in different situations.

Early scientists believed that heat was a physical substance. In the eighteenth century, they tried to prove the existence of an invisible fluid called "caloric." Although theories on caloric were disproved, scientists kept the word "calorie" as one of the units of heat, and we still use phrases such as "flow of heat."

Experiments showed that heat was not a substance, but a method of transferring energy. Basically, energy is the ability to make something move. The field of thermodynamics distinguishes between two general methods of transferring energy: work and heat. Work is a force that acts through displacement. Put more simply, work is a process equivalent to the lifting of a weight. This can be done through mechanical work, an electric generator, expanding gases (as in pistons), or other means.

Heat, as defined in thermodynamics, is the transferal of energy across the boundary of a system (or the surroundings) to another system at a lower temperature because of a difference in temperature between the two systems. In other words, heat is the transferal of energy resulting from the interaction of a hot and a cold object.

If you add an ice cube to a cup of coffee, heat will be transferred from the coffee to the ice cube. The ice will melt, and the mixture of coffee and water will reach the same temperature. This state is called thermal equilibrium. In thermodynamic terms, the mixture of coffee and water no longer contains heat, since there is no longer any transfer of energy from one system to another.

This sequence of photographs shows ice melting after it was placed in a hot cup of tea. Thermal equilibrium is achieved when the ice is fully melted.

Temperature is a measure of the intensity of heat, or "hotness." Long after the formulation of the first and second laws of thermodynamics, modern scientists were frustrated by the lack of a precise definition of temperature. In 1931, R. H. Fowler proposed the "zeroth law of thermodynamics" as a solution. It states that if two systems are in thermal equilibrium with a third system, the two systems will be in thermal equilibrium with each other. For example, a thermometer inserted in a beaker of

## In Hot Water

When you drink a cold liquid, your body provides heat to warm it. A hot liquid provides your body with extra heat. Either way, the liquid reaches thermal equilibrium in the body. Does this mean that you could survive on hot water alone? No, because the amount of additional heat is very small. You would have to drink about sixty-five cups of very hot (95°C or 203°F) water to equal the caloric content of one slice of bread!

water might read 25° Celsius. If another beaker of water is also measured at 25° Celsius, the two beakers of water are at thermal equilibrium. It might seem obvious, but the concept is crucial to temperature measurement in thermodynamics.

### Rumford's Cannons and Joule's Experiments

Two men provided the proof that eventually led to the demise of the caloric theory and the formulation of the first law of thermodynamics.

In 1797, the physicist Benjamin Thompson, also known as Count Rumford, was overseeing the manufacture of cannons. Rumford was fascinated with the phenomenon of heat. As an experiment, he bored the barrel of a cannon immersed in water with a dull drill bit turned by horses. Friction from the drill bit grinding the barrel produced heat. After two and a half hours, onlookers were astonished to see that the water

Count Rumford was the first scientist to propose that heat is a form of energy. He did so in 1798 in a published article called "Inquiry Concerning the Source of Heat Which Is Excited by Friction." Because of his work, he is known as the father of thermodynamics.

was boiling without being heated by a fire!

Rumford believed that he had proven that motion produced heat and that heat was not a substance called "caloric." Other scientists of the day were not convinced. Rumford could not quantitatively connect motion and heat. His theory did not explain how motion could cause heat to travel through empty space, such as the sun warming the earth through the atmosphere.

The caloric theory of heat reigned for another half century until a scientist named James Joule began studying heat and work. Unlike Rumford, Joule kept precise measurements during his experiments, done in the 1840s. He began with experiments on electromagnets and electric current. Joule devised an apparatus where a pair of weights passing over pulleys performed work that generated an electric current. The current produced heat. Joule could calculate the amount of heat produced by the work, a figure he called the "mechanical equivalent of heat."

In an experiment similar to Rumford's, Joule attached a thermometer to a paddle wheel immersed in water. The paddle wheel was turned by a descending weight. Joule was able to measure precisely the work done by the weight and the resulting temperature increase of the water.

Rumford and Joule both demonstrated that work could be converted into heat. Joule proved that a

This is the water friction apparatus, including the paddle wheel, that James Joule used to prove that heat is a form of energy in 1843. Joule also proved that energy cannot be created or destroyed, and that it only changes form. This concept is known as the conservation of energy.

fixed amount of work is required to generate one unit of heat. Energy, not heat, is conserved during the transformation. This conclusion refuted the caloric theory of heat, which assumed that heat itself was conserved in the form of "caloric." Joule's work became the foundation of the first law of thermodynamics.

Joule measured mechanical work in "foot-pounds" and heat in Btu, or british thermal units. He calculated that 772 foot-pounds of work would produce one Btu. Today, we know that the actual figure for the mechanical equivalent of heat is 778.26 foot-pounds per Btu. More often, scientists use the joule, abbreviated J, as the unit for measuring energy, work, and heat. One joule equals 0.239 calories.

## Implications of the First Law

The first law of thermodynamics leads to a number of corollaries. The most important is the statement that any change in the internal energy of a system is equal to the difference between the heat absorbed by the system and the work done by the system on its surroundings. Internal energy is the energy present in a system when it is stationary, and it exists independent of the motion of the system. According to the first law, if a system performs work, its internal energy will decrease. If the system absorbs heat, its internal energy will increase. The reverse can also occur—work can be performed on the system and heat transferred out of the system.

In some cases, work is performed and heat is transferred, but there is no change in the internal energy of the system. This is called a cyclic process. Work is performed on the system—for example, the weight that turned Joule's paddle wheel. Heat is transferred out of the system. The amount of energy transferred by heat equals the amount of energy transferred by work. Cyclic processes are very important in explaining heat engines—devices that convert heat into mechanical work.

Besides disproving the caloric theory, the first law of thermodynamics put an end to another dream of early scientists. Inventors had long attempted to devise a perpetual motion machine. In theory, a perpetual motion machine produces more energy than the amount required to keep the machine running. It would be an engine that could run on its own with no outside source of heat, fuel, or power, and still have energy to perform other tasks.

Nineteenth-century inventors attempted to design waterwheels in a contained pool that would pump the water back to the top of the waterwheel. The waterwheel would never stop turning, and spare energy could be diverted to do additional work. But the first law states that energy cannot be created out of nothing. A perpetual motion machine violates the law of conservation of energy.

# 2 The Second Law of Thermodynamics

The first law states that heat can be converted into work, and work into heat, but the first law does not address the direction of heat transfer. We know that a mug of hot coffee will naturally cool as it sits on a table. Heat is transferred from the coffee to its surroundings. But what law of nature prevents the coffee from instead drawing heat from the surroundings? Why can't it spontaneously grow hotter and eventually boil, rather than cooling? (In scientific terms, a spontaneous event is one that occurs naturally without external causes.) The second law of thermodynamics addresses the distribution of heat and energy. One of its statements is that heat naturally flows from a hot object into a cooler object, not the other way around.

The second law of thermodynamics also has a profound practical impact on the theory and design of heat engines. A car engine, for example, burns gasoline to produce heat. The heat is transformed into the mechanical energy that runs the car. According to the second law, the engine cannot have

100 percent efficiency. It is impossible for all of the heat to be transformed into mechanical energy.

Finally, the second law addresses the topic of entropy. In everyday speech, "entropy" is often used as a synonym for "disorder" or "randomness." If you drop a cube of sugar into a mug of coffee, it will not keep its shape. The granules will dissolve and disperse throughout the liquid. In thermodynamic terms, the tendency toward a

This photo shows bubbles of steam in boiling water. The rising bubbles of steam are water vapors escaping. When water boils, it is undergoing a transition from liquid to gas by transferring heat into the air.

greater degree of energy dispersal is called an increase in entropy. Scientists can quantify entropy and use an understanding of entropy to predict the behavior of objects in thermodynamic systems.

According to the second law, any spontaneous process in an isolated system will lead to an increase in the entropy of the system. It is possible for entropy to decrease in nonisolated systems. But overall, the number of natural processes in which entropy increases outnumbers those in which it remains constant or decreases. When all systems are taken together, natural processes are accompanied by an increase in the entropy of the universe.

## The Carnot Engine

The second law of thermodynamics will forever be associated with the work of a single man. Sadi Carnot was a French military engineer who dedicated his short life to studying the steam engine. In 1824, Carnot published a short book called *Reflections on the Motive Power of Heat and on Machines Appropriate for Developing This Power.* He described a theoretical engine, now called a Carnot engine, that operated at the ideal efficiency. Because of friction and other factors that cause engines to lose energy, no real engine can operate as efficiently as the Carnot engine.

Sadi Carnot is often referred to as the founder of modern thermodynamics because of his theory that set the groundwork for the formulation of the second law of thermodynamics.

Carnot was the first scientist to understand the potential of the cyclic process in creating an efficient engine. His engine operated in an ideal reversible cycle called the Carnot cycle. During Carnot's day, steam engines operated at only about a 6 percent efficiency. Ninety-four percent of the coal burnt to fuel the engine was wasted.

Some engineers tried to improve efficiency by limiting heat loss in engines. Carnot took a different approach. He realized that it was impossible to transform heat into work without discarding some heat. Carnot viewed the engine as having two outlets: one for work and one for leftover heat. He used a waterwheel as an analogy. Just as water moves from the top of the waterwheel to a pool at the bottom, heat in his engine flowed from a hot source to a cold sink.

Carnot considered what could result if his engine were run backward. Such a machine would consume work, remove heat from a cold area, and discard heat into already warm surroundings. In short, this describes a refrigerator.

There was one great flaw in Carnot's work. Carnot based his studies on the caloric theory of heat. He published his *Reflections*, which served as the foundation for the second law, before Joule performed his experiments on work and heat. At the middle of the nineteenth century, scientists could not see how both Carnot and Joule could be correct.

Two physicists reconciled Carnot's work with Joule's findings on the nature of heat. In doing so, they put forth two restatements of the second law of thermodynamics and revealed further implications of the second law.

## Formulating the Second Law

A physicist named William Thomson, usually called Lord Kelvin, formulated one statement of the second law. According to the Kelvin statement (as stated in *The Second Law*, by P. W. Atkins), no process is possible in which the sole result is the absorption of heat from a reservoir and its complete conversion into work. In other words, when heat is converted into work, it is impossible for 100 percent of the heat to be converted into work. Some heat will be lost.

This implies the impossibility of what is called a "perpetual motion machine of the second kind." Such a machine could convert heat from nature into unlimited quantities of work. The second law establishes that it would be impossible to build an engine that could completely convert heat from, say, the atmosphere or a lake, into work.

Another physicist, Rudolph Clausius, formulated another statement of the second law in a monograph. It was titled "On the Motive Power of Heat and the Laws Which Can Be Deduced from It for the Theory of Heat." Published in 1850, the paper defined both the first and second laws and marked the creation of the science of thermodynamics. According to the Clausius statement of the second law (as stated by Atkins), no process is possible in which the sole result is the transfer of energy from a cooler to a hotter body. Heat flows naturally from hot to cold areas, but not the reverse. A cup of coffee will never spontaneously grow hotter and boil.

The Clausius statement rules out the possibility of spontaneous refrigeration. If heat could flow naturally from a hot area to a cold area, a freezer could

Refrigerators take heat from a cold place and transfer it to a warmer place.

Rudolph Clausius is the German mathematician and physicist who is credited with making thermodynamics a science. Entropy was his most important theory.

be invented that could run with no need of energy from electricity. But like perpetual motion machines, this would violate the laws of thermodynamics.

Clausius also invented the term "entropy," based on the Greek word for "transformation." During his studies in the new field of thermodynamics, he focused on a certain mathematical ratio: heat over temperature. Mathematically, it is written Q/T, since Q is the symbol used for heat in thermodynamic equations. The ratio represents the amount of heat transferred and the temperature, before and after the heat transfer. Clausius examined the ratio for different kinds of processes involving heat transfer, beginning with the heat transferred in a reversible engine. The ratio stayed constant for a reversible cycle. Clausius went on to examine irreversible processes, such as heat being transferred from a hot cup of coffee. Q/T increased

## Patenting a Perpetual Motion Machine

The U.S. Patent Office has a policy of refusing to grant patents for any type of perpetual motion machine on the grounds that it would violate the laws of physics. In 1972, one person did manage to slip past this rule. U.S. Patent No. 3,670,500 is for a "thermodynamic power system" that violates the second law of thermodynamics.

for such processes. Clausius renamed the Q/T ratio, calling it entropy. Unlike energy, entropy is not conserved during natural processes. Clausius concluded that the entropy of the universe naturally tends to increase.

## An Example of Discarded Heat: Thermal Pollution

For years, environmentalists have attacked air pollution emitted by fossil fuel-burning power plants. Most people are unaware of another by-product of these plants: thermal pollution. As shown by the second law, it is impossible to convert all of the heat generated in a power plant into mechanical energy. Waste heat is released into the environment.

Power plants burn fossil fuels to produce heat, which converts water into steam. At this point, about 12 percent of the energy is lost in the chimney stack. The steam passes through a turbine that generates electrical power from the steam. The leftover steam, which is now waste heat, is transferred into a condenser. It condenses into

Power plants are usually located by a body of water, such as a river, pond, or lake. The body of water serves as a cooling reservoir for the plant. This photograph shows the cooling towers of the Three Mile Island power plant on the Susquehanna River in Harrisburg, Pennsylvania.

water, which is usually released into a river, pond, or lake.

According to mathematical equations based on the second law, fossil fuel power plants can have a maximum efficiency of only about 66 percent. In reality, plants generally have an overall efficiency of about 40 percent. About 60 percent of all heat generated is released into the environment.

# 3 Heat Engines and Refrigerators

Today, in the twenty-first century, we view the steam engine as a technologically primitive relic of the past. We travel by jet and derive electricity from nuclear fuel. But jet engines and nuclear power plants adhere to the fundamental principles established in the nineteenth century by the second law of thermodynamics.

Carnot dreamed of creating a maximum efficiency engine, an ambition shared by engineers today. He used thermodynamic concepts to solve practical engineering issues. His engine operates in an ideal, reversible cycle, between a high and a low temperature reservoir. The efficiency of the engine depends on the temperatures of the hot and cold reservoirs. One only needs to know the two temperatures to calculate the efficiency. The mathematical term for this quantity is the Carnot efficiency or the Carnot factor. Carnot's theorem states that no real engine of the same type as the theoretical Carnot engine can be more efficient than the Carnot engine.

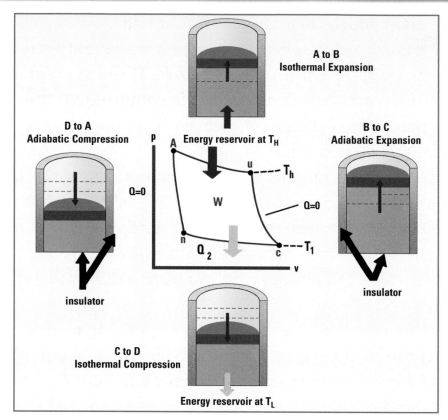

This illustration shows the four stages of the Carnot cycle: isothermal expansion, adiabatic expansion, isothermal compression, and adiabatic compression. The Carnot cycle is a theoretical ideal, which establishes the maximum efficiency of a heat engine. It assumes that the processes involved in a heat engine are reversible and that there is no change in entropy, conditions which do not exist in real life.

## Inside the Carnot Engine

An understanding of the Carnot engine requires some basic knowledge of gases. When a gas is compressed into a smaller volume container, such as by a piston, its pressure increases. The pressure of a gas also increases with temperature.

If a gas is kept in contact with a heat source while it is compressed, its temperature will not change. This is called an isothermal compression.

The heat source is usually called a heat reservoir. For example, a reservoir could be a water bath that remains at a constant temperature. If there are no heat interactions when the gas is compressed, the temperature of the gas increases. This is called an adiabatic compression.

In a Carnot engine, a quantity of gas is contained in a cylinder fitted with a piston. It exchanges heat with two reservoirs, one at a high temperature and one at a low temperature. The Carnot cycle consists of four reversible processes, two isothermal and two adiabatic.

During the first step, the gas is put into contact with the high temperature reservoir. The gas absorbs heat from the reservoir and expands in volume—an isothermal expansion, since the temperature of the gas is constant. This step produces work by raising the piston.

The second step is an adiabatic expansion. The high temperature reservoir is removed and replaced by an insulator. The gas expands adiabatically as pressure and temperature fall. This step also produces work by raising the piston.

No further work can be done by the gas beyond this point, so the gas must be returned to its initial condition. When the gas is cool, it is put on the low temperature reservoir for the third step, an isothermal compression. The piston begins to compress the gas back to its original volume. The gas tends to heat as it is compressed, but the low temperature reservoir absorbs the extra heat. This step consumes work since the piston performs work on the gas.

The fourth step is an adiabatic compression. The low temperature reservoir is removed and replaced by an insulator. The piston continues to compress the gas until it has returned to its original temperature and pressure. This step also consumes work. The gas has returned to its initial conditions, and the cycle can begin once again.

The Carnot cycle serves as an example of the statement of the second law dictating that not all heat can be converted into work. An engine operating on a cycle must have a low temperature reservoir, and some heat must be discarded into that reservoir. In an automobile, for example, extra heat is discarded through the radiator and exhaust pipe.

## Maximizing Engine Efficiency

At first glance, the steps of the Carnot cycle may seem to cancel each other out. The gas produces work during the first two steps, but the last two steps both consume work. How does the engine produce net work?

The answer lies in the nature of gas. Less work is required to compress gas when it is cool. In the Carnot engine, gas is cooled before it is compressed by the piston. The net work produced depends on the difference between the temperatures of the high and low temperature reservoirs. Therefore, the greater the temperature difference between the two reservoirs, the greater the amount of work produced.

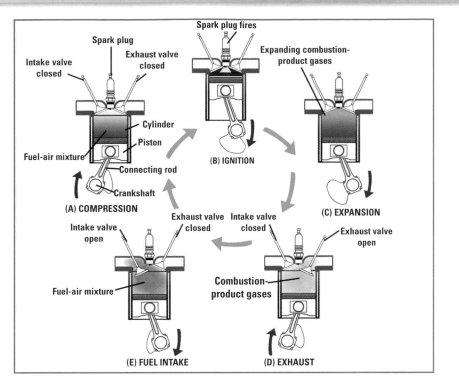

This illustration shows the complete work cycle of a gasoline engine. During compression, the piston adiabatically compresses the mixture of fuel and air in the cylinder. Combustion occurs when maximum compression is reached. Combustion increases the internal energy of the gas, causing the gas to expand and the piston to do work on the crankshaft. The work of the piston transfers heat through the wall of the cylinder and hot exhaust gases from the cylinder. Finally, as the piston moves downward once it finishes its work, it pulls a new fuel-air mixture into the cylinder to restart the cycle.

The ratio of the work produced to the heat absorbed is the efficiency of the cycle. Carnot showed that the efficiency depends only on the temperature of the two reservoirs:

$$e_c = \frac{T_H - T_L}{T_H}$$

In this equation, $e_c$ is the efficiency of the Carnot engine. $T_H$ and $T_L$ are the temperatures, measured in

Kelvin, of the high and low temperature reservoirs. No real engine operating in the cycle between the temperatures $T_H$ and $T_L$ can be more efficient than the theoretical Carnot engine.

It might seem that to maximize efficiency, the best solution would be to maximize the difference between the two temperatures. In reality, however, it is not practical to make the high temperature reservoir so hot that it could damage other components of the engine. The low temperature reservoir is usually around room temperature.

## Reversible and Irreversible Processes

Carnot's theoretical engine includes an ideal, reversible cycle that is impossible in the real world. A reversible process is one that, when it is completed, has returned exactly to its initial conditions. There are no processes in the natural world, however, that are truly reversible.

A ball dropped onto the floor will bounce a couple times, roll away, and come to rest. It will never spontaneously reverse its path, roll toward you, and bounce into your hand. An ice cube left in a glass on a warm day will melt. That particular ice cube will never re-form.

Some processes are almost reversible. Parts of an engine may move so slowly that they are always in near equilibrium, a state of no motion or change. Such a process would move so slowly that it would take an infinite amount of time to complete. When

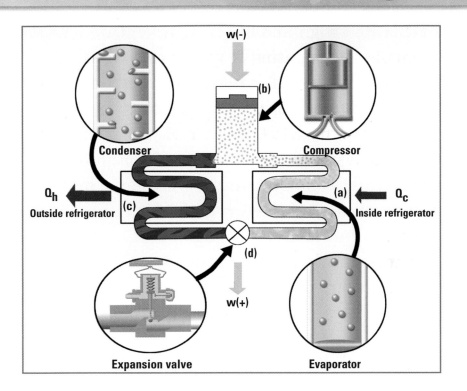

w(-)

(b)

Condenser

Compressor

$Q_h$

Outside refrigerator

(c)

(a)

$Q_c$

Inside refrigerator

(d)

w(+)

**Expansion valve**

**Evaporator**

Cooling in most refrigeration systems is achieved by the evaporation of a liquid refrigerant under reduced pressure and temperature. This illustration shows the four steps of a typical refrigeration cycle. The refrigerant moves from the evaporator to the compressor to the condenser and through an expansion valve back to the evaporator. At each stage, energy is transferred to and from the refrigerant, which changes form between liquid and vapor, by either heat or work.

scientists write of reversible processes, they are generally talking about them in theoretical terms. A real engine cannot take an infinite amount of time per cycle. Components in a real engine, unlike in the Carnot engine, lose energy through friction. A breeze might cool one of the reservoirs very slightly.

## Heat Pumps

Carnot speculated that his engine could operate backward. Rather than transfer heat from a high temperature reservoir to a low temperature reservoir, it would do the opposite. It would remove heat from a low temperature reservoir. According to the second law of thermodynamics, such a process could not be spontaneous. Therefore, instead of producing net work, as a heat engine does, it would require an input of work. A general term for this type of machine is a heat pump. Refrigerators and air conditioners are two kinds of heat pumps.

In designing heat engines, we want to maximize the temperature differences of the reservoirs to operate at the highest efficiency possible. A heat pump is a different matter. A large difference in temperature between the reservoirs means that a large amount of energy is required to operate the heat pump. It is more desirable that a heat pump minimize the temperature difference between reservoirs and use a minimal amount of energy.

## Cooling Down

It's a hot day. You open the fridge door and suddenly feel better as you feel the icy air on your skin. And the entire room's a bit cooler now that you've let some of the cold air into the kitchen, right?

Not quite. A refrigerator removes hot air from its interior and expels it into the surroundings. If you leave the door open, the refrigerator only moves heat from one part of the room to another.

Engineers try to design "energy efficient" refrigerators that do not use much electricity.

The effectiveness of a heat pump is described as the "coefficient of performance," or the C.O.P. Just as there is an equation for the ideal Carnot efficiency, there is an equation for the C.O.P. of the ideal reversible heat pump:

$$C.O.P. = \frac{T_L}{T_H - T_L}$$

T is measured in Kelvin.

If you compare the equations, you will see that both depend on the difference between temperatures of the two reservoirs. The Carnot efficiency increases as the temperature difference increases. The opposite occurs for the C.O.P.: the effectiveness of the heat pump decreases as the temperature difference increases.

# 4

# Understanding Entropy

Clausius coined the term "entropy," defining it in terms of heat and temperature. He went further in stating that the entropy of the universe tends toward a maximum. But Clausius did not analyze the meaning of the existence of entropy. What exactly is entropy? It is a property of matter, since it can be measured in a laboratory, but it is not something we can see or touch.

## The Nature of Entropy

The word "heat" has a more specific meaning in thermodynamic terms than in everyday usage. When we speak of "heat" in our daily lives, we might mean either warmth or temperature—not necessarily the scientific definition of a transfer of energy from a higher to a lower temperature. In the same way, entropy has taken on a significance in general English other than its original meaning in thermo-dynamics. "Entropy" is sometimes used synonymously with "chaos" or "disorder." In thermodynamics, though, entropy always refers to the measure of the dispersal of energy, at a certain temperature. A

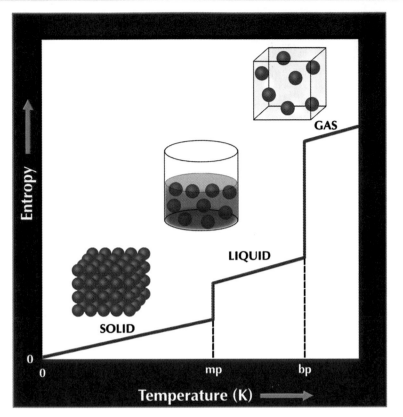

This diagram demonstrates changes in entropy as a result of changes in temperature. As the temperature increases and matter changes from solid to liquid to gas, the configuration of the matter's molecules changes from being tightly packed to loosely arranged, until they begin to diffuse.

pan heated on the stove will disperse heat throughout the entire kitchen. Entropy measures the dispersal of that energy as it moves outward from a concentrated area. Shuffling a new deck of cards is sometimes used as an analogy for entropy increase, but this is a misleading example. Although the cards move from an ordered state to a random state, there is no energy change in the deck of cards itself.

Entropy originates in the microscopic arrangement of matter, which exists as solids, liquids, or gases. Matter is made up of atoms, usually combined with other atoms to form clusters called molecules. Molecules are in constant motion, colliding with each other and changing direction. Nineteenth-century scientists compared the motion of gas molecules to a game of billiards, with one ball ricocheting off another. In this analogy, every ball is constantly in motion, moving at very high speeds.

Physicists later linked entropy and probability. If you drop a dozen billiard balls on a table, there are millions of different arrangements in which they could come to rest. There is the chance that they could form, for instance, four perfect rows. It is much more probable that they will land in a random pattern, since there are many more possible random patterns than orderly patterns. A neat arrangement of billiard balls, if disturbed, will likely result in a random pattern. A random pattern, however, will also probably result in another random pattern. Gas molecules ricocheting off each other in space are unlikely to spontaneously form an orderly arrangement.

What does this have to do with entropy, as defined as the measure of the dispersal of energy? As molecules collide, they exchange energy. Unless it is an isolated system, an object exchanges energy with the surroundings as well as with other molecules within the system. If the object is heated, such as our pan on the flame, its energy increases. The molecules that make up the pan begin to move more rapidly. They disperse this extra energy from heat to the molecules in their surroundings, which are moving more slowly. Overall entropy increases during the process.

A change in entropy can occur without a change in temperature. Phase changes are the transition from one state of matter—solid, liquid, gas—to another. Molecules in solids are tightly held in a rigid configuration and can move only slightly. Molecules in liquids can move more freely, but they still interact with each other. Gas molecules move freely and tend to diffuse. During phase changes, entropy changes even though the temperature remains constant.

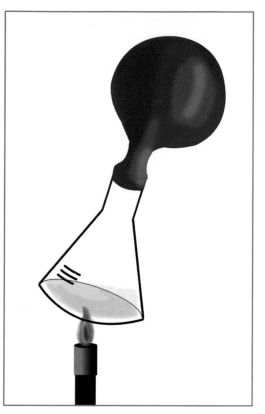

Water turns into steam by heat energy. As the temperature increases, the quicker the water/steam molecules move around. The speeding molecules hit against the surface of the balloon, increasing the pressure against the surface. This causes the balloon to expand.

If you add heat to a block of ice, it melts at 0°C and vaporizes at 100°C. At 0° and 100°, entropy increases as the water goes through a phase change, from solid to liquid and liquid to gas. This is because the dispersal of energy is increasing during the phase changes. A liquid is a more widely dispersed form of matter than a solid, just as a gas is more widely dispersed than a liquid.

Phase changes are just one example of how entropy can change without being accompanied by a temperature change. Entropy tends to increase in any case that molecules can freely move to disperse themselves throughout a larger area.

## Degradation of Energy

According to the second law of thermodynamics, work can be completely transformed into heat, but the reverse can never be true. Some heat will always be discarded in the transformation of heat to work. For this reason, the second law is sometimes called the law of degradation of energy.

Energy dumped into the low temperature reservoir of a heat engine can be described as lower quality than energy in the high temperature reservoir.

## The Arrow of Time

Entropy is sometimes discussed in the context of "the arrow of time." Every event in nature could conceptually occur in reverse, but in reality, natural processes are irreversible. Perfume dispersed into the air will not spontaneously flow back into the bottle. Events proceed in the direction favored by entropy, and entropy is constantly increasing. In this way, entropy gives direction to the flow of time.

Since it is less useful for doing work, the quality of the energy in the low temperature reservoir has been degraded. In general, for processes involving heat transfer, the amount of energy available for doing work decreases.

At the same time, entropy increases during heat transfers. These two tendencies can be taken together: as entropy increases, energy is constantly being degraded and stored at lower temperatures. Natural processes tend to generate entropy and diminish the quality of energy.

Some physicists have theorized that the universe could eventually reach a state of maximum entropy. The universe would be in a state of thermal equilibrium, with all energy uniformly distributed. This scenario is known as the "heat death" of the universe.

## The Third Law of Thermodynamics

The third law of thermodynamics addresses a practical aspect of entropy. We know that if a beaker of water is heated, its entropy increases. Scientists can measure the entropy change in a lab. But measuring the change in entropy tells us only about difference in entropy between the initial and the final state. This measurement is relative. It conveys nothing about the absolute entropy of the beaker of water at a specific temperature.

In 1906, a German chemist named Walther Nernst presented the Nernst heat theorem. This law is often referred to as the third law of thermodynamics. It states that at absolute zero, a perfectly crystalline substance has zero entropy.

Absolute zero is the lowest theoretical temperature possible, representing a complete absence of heat. It is equal to 0 on the Kelvin temperature scale, -273.15°C, or -459.67°F. Absolute zero cannot be achieved in a finite number of steps. In other words, no matter how cold something is, it can always be made colder without ever reaching

Walther Nernst *(right)* won the 1920 Nobel Prize for Chemistry in recognition of his formulation of the third law of thermodynamics.

absolute zero. Lab experiments studying the behavior of matter at absolute zero can only be performed at temperatures near absolute zero.

At absolute zero, molecular motion comes to a standstill. There are no energy interactions between molecules. Therefore, a perfectly crystalline substance has zero entropy at absolute zero. A perfectly crystalline substance is one that has all molecules and atoms perfectly aligned with each other. Diamond is one example of a perfectly crystalline substance. If the molecules and atoms are not perfectly aligned, there will be interactions between them. The entropy of such a substance will be greater than zero, even at absolute zero.

The third law of thermodynamics establishes a reference point for the entropies of all substances. Based on the third law, scientists can determine the entropy of any element or compound. These molar entropies, as they are called, state the entropy acquired as the substance is heated from absolute zero to room temperature. Molar entropies are an important tool in understanding chemical reactions and phase changes.

# Glossary

**absolute (AB-suh-LOOT)** Independent of arbitrary standards of measurement.

**beaker (BEE-ker)** A type of laboratory glassware consisting of a cylindrical cup with a notch on the top to allow for the pouring of liquids.

**boundary (BOWN-duh-ree)** The line or plane indicating the limit or extent of something.

**British thermal unit (BRIH-tish THER-mul YOO-nit)** A unit of heat equal to the amount of heat needed to raise one pound of water 1°F (–17.22°C); usually denoted as BTU.

**calorie (KA-luh-ree)** The quantity of heat required to raise the temperature of one gram of water by 33.8°F (1°C).

**compound (KOM-pownd)** A material formed from elements chemically combined in definite proportions by mass.

**displacement (dis-PLAYS-mint)** The act of physically moving something out of position.

**electromagnet (ih-lek-tro-MAG-neht)** A temporary magnet made by coiling wire around an iron core. When current flows in the coil, the iron becomes a magnet.

**foot-pound (FUT-pownd)** A unit of work equal to a force of one pound moving through a distance of one foot.

**heat (HEET)** The transferal of energy across the boundary of a system (or the surroundings) to another system at a lower temperature because of a difference in temperature between the two systems.

**insulator (in-suh-LAY-tur)** An object designed to prevent the loss of heat or electricity.

**joule (JOOL)**  A unit of energy used to calculate the amount of work done within a certain space.

**monograph (mä-ne-graf)**  A scholarly book, article, or pamphlet on a specialized subject.

**plasma (PLAZ-muh)**  A gaslike state of matter consisting of positively charged ions, free electrons, and neutral particles.

**system (SIS-tim)**  In thermodynamics, the part of the universe of interest or under scrutiny.

**thermal equilibrium (THER-mul ih-KWIH-lih-brih-um)**  Point at which energy gained by a system is balanced by the energy lost.

**thermodynamics (THER-mul-DY-nuh-miks)**  The study of energy transfers and transformations.

**work (WURK)**  The energy required to move an object against an opposing force.

# For More Information

American Chemical Society
1155 Sixteenth Street NW
Washington, DC 20036
(800) 227-5558
Web site: http://www.chemistry.org/portal/a/c/s/1/home.html

American Institute of Physics Center for History of Physics
One Physics Ellipse
College Park, MD 20740-3843
(301) 209-3100
Web site: http://www.aip.org/history

The International Union of Pure and Applied Chemistry
IUPAC Secretariat
P.O. Box 13757
Research Triangle Park, NC 27709-3757
(919) 485-8700
Web site: http://www.iupac.org/dhtml_home.html

## Web Sites

Due to the changing nature of Internet links, the Rosen
Publishing Group, Inc., has developed an online list of Web
sites related to the subject of this book. This site is updated
regularly. Please use this link to access the list:

http://www.rosenlinks.com/liph/lath

# For Further Reading

Fleisher, Paul. *Matter and Energy: Principles of Matter and Thermodynamics.* Minneapolis: Lerner Publications Company, 2001.

Greenberg, Arthur. *A Chemical History Tour: Picturing Chemistry from Alchemy to Modern Molecular Science.* New York: Wiley-Interscience, 2000.

Gribbin, John. *The Scientists: A History of Science Told Through the Lives of Its Greatest Inventors.* New York: Random House, 2003.

Herr, Norman, and James Cunningham. *Hands-On Chemistry Activities with Real Life Applications, Volume 2.* San Francisco: Jossey-Bass, 2002.

# Bibliography

Atkins, P. W. *The Second Law.* New York: Scientific American Library, 1984.

Craig, Norman C. *Entropy Analysis: An Introduction to Chemical Thermodynamics.* New York: VCH Publishers, Inc., 1992.

Fenn, John B. *Engines, Energy, and Entropy: A Thermodynamics Primer.* San Francisco: W. H. Freeman and Company, 1982.

Goldstein, Martin, and Inge F. Goldstein. *The Refrigerator and the Universe: Understanding the Laws of Energy.* Cambridge, MA: Harvard University Press, 1993.

Kondepudi, Dilip, and Ilya Prigogine. *Modern Thermodynamics: From Heat Engines to Dissipative Structures.* New York: John Wiley and Sons, Inc., 1998.

Lambert, Frank. "Entropy Sites—A Guide." April 2004. Retrieved May 11, 2004 (http://www.entropysite.com).

Van Wylen, Gordon J., and Richard E. Sonntag. *Fundamentals of Classical Thermodynamics.* New York: John Wiley & Sons, Inc., 1965.

Von Baeyer, Hans Christian. *Maxwell's Demon: Why Warmth Disperses and Time Passes.* New York: Random House, 1998.

# Index

## About the Author

Rose McCarthy is a freelance writer who lives in Chicago, Illinois.

## Photo Credits

Cover © Francoise Sauze/Science Photo Library; p. 5 © Stuart Westmorand/Corbis; p. 8 © Sam Ogden/Photo Researchers, Inc.; p. 10 by Tahara Anderson; pp. 12, 22 © Science Photo Library; p. 13 © Science Museum, London/Topham-HIP/The Image Works; p. 17 © Alfred Pasieka/Science Photo Library; p. 18 © J-L Charmet/Science Photo Library; p. 21 © Sally A. Morgan; Ecoscene/Corbis; p. 24 © W. Cody/Corbis; pp. 26, 29, 31, 35, 37 by Geri Fletcher; p. 40 © Bettmann/Corbis.

**Designer:** Tahara Anderson; **Editor:** Wayne Anderson